MW00800132

Copyright © 2024 Dr. Jessica Lawson

All rights reserved. No part of this book may be reproduced,

distributed, or transmitted in any means,

Including photocopying, recording, or other electronic

or Mechanical methods, without the prior written consent of the

copyright owner, except in the case of brief quotation

embodied in critical articles, review, and certain other

noncommercial uses permitted by copyright law.

ISBN: 9798991140812

# This book belongs to:

*Come one,* come all, and listen very well! To the story, the magic, the legend, the tale, of a bright young star whose name is quite simple and how one sad day she lost her twinkle.

*Nora* was a real star, not like the ones on TV. She had *"STAR POWER"*, but not like in the movies. Nora lived far away, high in the sky you see. Above the clouds in a far away galaxy.

Everywhere Nora went with other stars she would mingle. She made everyone laugh with a wink and a wiggle.

Have you seen my twinkle?

No, i'm sorry...

Then one day the other stars noticed a change. Nora had lost her twinkle and that was quite strange.

What good is a star if she cannot shine bright? What's the use if she cannot sparkle in the night?

Just the thought of life without shining, made Nora feel sad and want to run away crying.

But Nora was not going to give up so
easily. She packed up her things and
began a journey

To find what she lost and
make things right; to get
back her twinkle
and share
her light.

Nora went searching and landed on the moon. She looked under moon rocks all afternoon.

Nora looked up, down and all around but her twinkle was nowhere to be found.

Nora traveled further and found herself with Saturn. She danced on his rings as they both erupted with laughter.

Nora asked Saturn to help her find her light. They looked left and right, but it was nowhere in sight.

It's not here.

Okay. thank you for helping.

Nora waved goodbye to Saturn and wished she could stay, but she had to keep searching. The next stop was the Milky Way.

The Milky Way's colors were so inviting that little Nora's heart gasped. She almost forgot her mission, but quickly remembered her task.

"Have you seen my twinkle?" Nora asked the Milky Way and listened carefully to what she had to say. "No," Milky Way answered. "But you might find your twinkle on Planet Number Nine. If you're quick you might make it in time."

Nora grew sad and thought all hope was
lost. Who knew finding her twinkle
would come at such a cost?
Planet Nine was so very far away!
Nora knew she'd best be on her way.

Nora floated for what seemed like forever. Then suddenly, there was a change in the weather. The wind started to blow, and she knew it was a sign. Just over the horizon was Planet Number Nine!

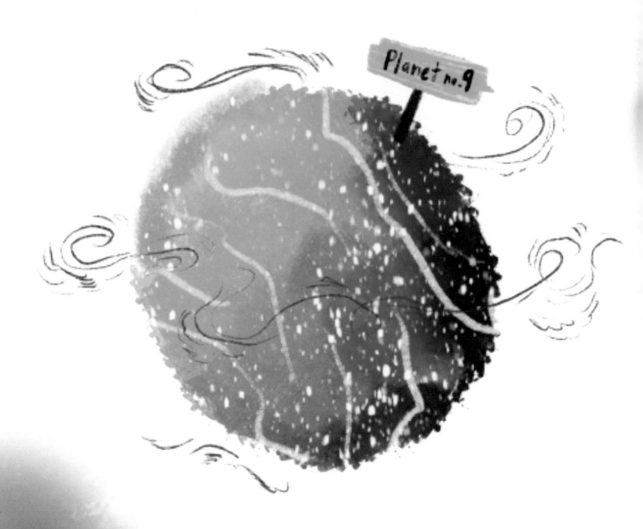

Nora tried to look around but there was barely any light. She heard so many strange noises, but she could see no one in sight. The air was so cold it gave her quite a chill. Then poor Nora started to feel quite ill.

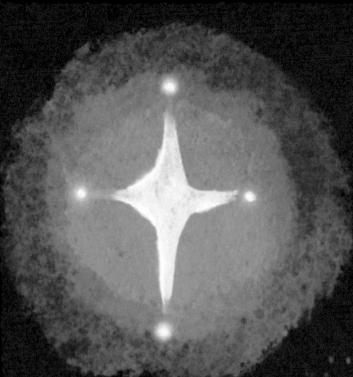

Then out of the darkness,
Nora saw a tiny flicker.
As it got brighter and
brighter,

Nora ran quicker and
quicker! She was running as
fast as her little legs would
go. It was her twinkle!
Nora's heart started to glow.

Nora was so happy to be on Planet Number Nine
Until she heard a voice say,

"Stop! That twinkle is mine!"

Nora was confused. Who would make such a claim?
She thought to herself,
"That's what I was about to say."

A mysterious figure appeared through the haze
and a tiny voice whispered,
"Hello, my name is Grace."
It was at this moment,
Nora knew she had to make a choice,
Or she would not only lose twinkle…
but she'd also lose her voice.

"Hello Grace," said Nora. "How do you do?
My name is Nora and it's nice to meet you.
Grace, I'm afraid you're mistaken," she gathered the
courage to say,
"That is my twinkle. Why did you take it away?"

Grace sadly answered, "It's dark and lonely here, I heard your twinkle was special and brought everyone such cheer. I was only going to borrow it, so things wouldn't be so black, But once I had it in my grasp, I just couldn't give it back."

Grace continued, "I'm sorry I took your twinkle, I know what I did was wrong. I'll give it back and when It's dark, I'll try my best to be strong. Would you please forgive me? I know I caused you pain. I'll give back your twinkle and I'll never take it again."

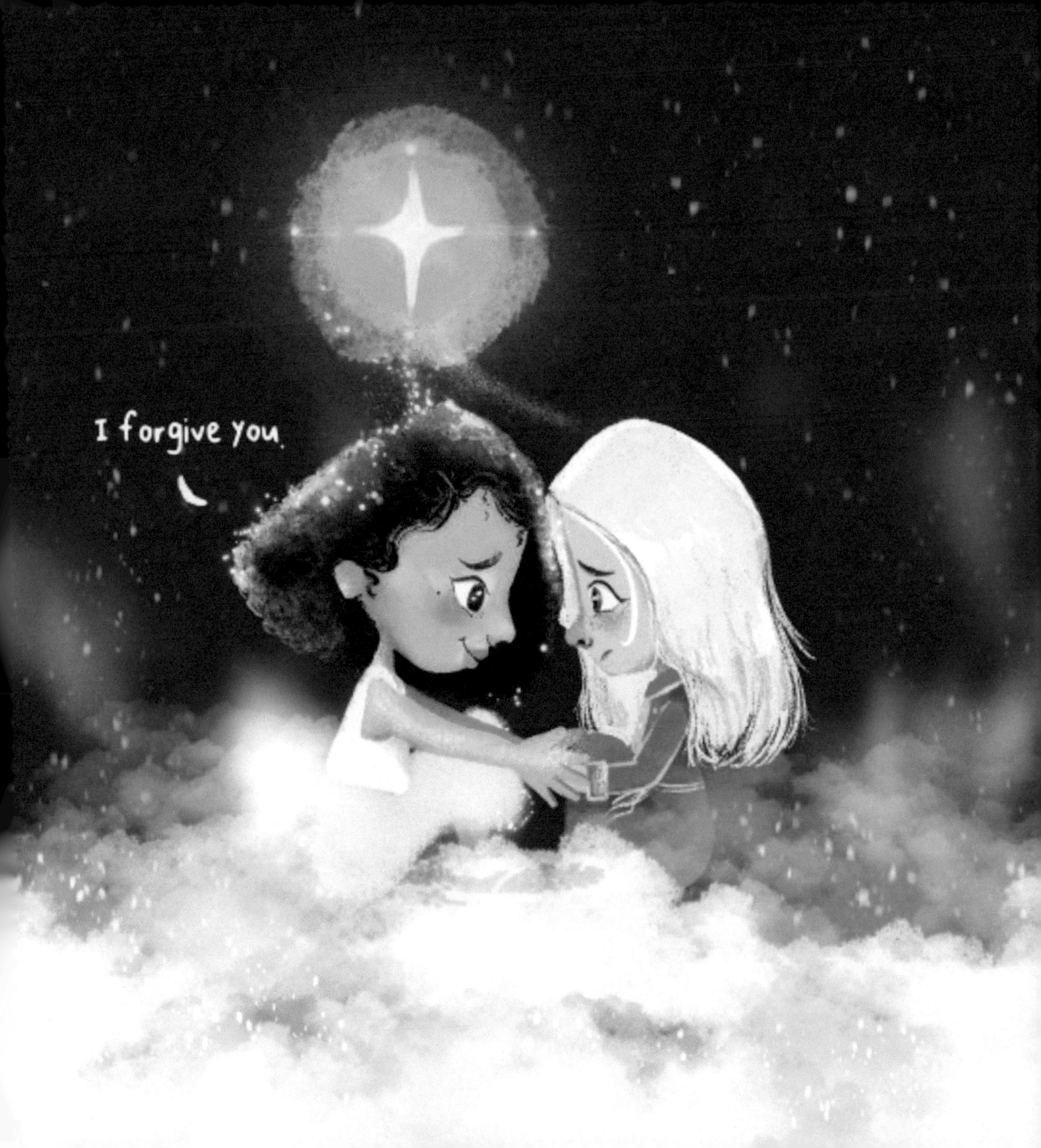
I forgive you.

Nora was happy! She felt very excited indeed, but couldn't help feeling just a little bit worried.

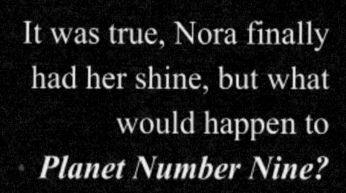

It was true, Nora finally had her shine, but what would happen to **Planet Number Nine?**

Nora remembered the joy she shared even when her twinkle was lost. She thought of the good times with all her friends whose paths she'd crossed.

Without her twinkle, Nora found joy in every day. Could it be Nora's twinkle was just hidden and never really went away?

Nora took a deep breath and made a tough decision. Leaving without her twinkle was not what she envisioned.

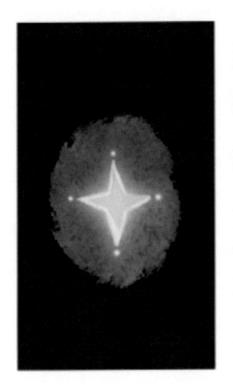  

But if sharing her light would make Grace's life brighter, she'd be happy to do something to make Grace feel a little nicer.

So, Nora looked within herself and with all her might, shined a glimmer of her twinkle, and she shared her light!

Now, if Grace felt sad in the darkness, and things didn't feel quite right, she could look inside herself and shine her very own light.

Nora said, "Farewell" and Grace waved goodbye. So much joy in their hearts and a sparkle in both their eyes.

Planet no. 9

*As Nora headed home,* she knew she would always remember this day. She learned the light living inside her was something no one could ever take away.

# ABOUT THE AUTHOR

Born under the vast starry skies of El Dorado, AR, Dr. Jessica Lawson has always been enchanted by the wonders of the universe. A passionate storyteller and adventurer at heart, Jessica channels her love for travel, exploration, and the marvels of science into every tale she weaves. Academically trained as a pharmacist, Jessica dedicates her days to improving healthcare and promoting health equality for all. In her free time, she brings to life magical stories that ignite the imagination and inspire others to explore and unleash their own creativity.

Through *The Star That Lost Its Twinkle,* Jessica aims to inspire those everywhere to find their inner strength, dream impossible dreams, and embrace the journey to finding one's own voice. Join Jessica on this whimsical journey and discover the twinkle within Yourself, one star at a time.

Made in the USA
Columbia, SC
24 September 2024

42558436R00024